~A BINGO BOOK~

Our Planet Earth Bingo Book

COMPLETE BINGO GAME IN A BOOK

Earth as Seen From *Apollo 17*
SOURCE: NASA

Written By Rebecca Stark
Educational Books 'n' Bingo

ISBN 978-0-87386-439-8

Educational Books 'n' Bingo

Printed in the U.S.A.

OUR PLANET EARTH BINGO DIRECTIONS

INCLUDED:

List of Terms

Templates for Additional Terms and Clues

2 Clues per Term

30 Unique Bingo Cards

Markers

1. **Either cut apart the book or make copies of ALL the sheets. You might want to make an extra copy of the clue sheets to use for introduction and review. Keep the sheets in an envelope for easy reuse.**

2. Cut apart the call cards with terms and clues.

3. Pass out one bingo card per student. There are enough for a class of 30.

4. Pass out markers. You may cut apart the markers included in this book or use any other small items of your choice.

5. Decide whether or not you will require the entire card to be filled. Requiring the entire card to be filled provides a better review. However, if you have a short time to fill, you may prefer to have them do the just the border or some other format. Tell the class before you begin what is required.

6. There are 50 terms. Read the list before you begin. If there are any terms that have not been covered in class, you may want to read to the students the term and clues before you begin.

7. There is a blank space in the middle of each card. You can instruct the students to use it as a free space or you can write in answers to cover terms not included. Of course, in this case you would create your own clues. (Templates provided.)

8. Shuffle the cards and place them in a pile. Two or three clues are provided for each term. If you plan to play the game with the same group more than once, you might want to choose a different clue for each game. If not, you may choose to use more than one clue.

9. Be sure to keep the cards you have used for the present game in a separate pile. When a student calls, "Bingo," he or she will have to verify that the correct answers are on his or her card AND that the markers were placed in response to the proper questions. Pull out the cards that are on the student's card keeping them in the order they were used in the game. Read each clue as it was given and ask the student to identify the correct answer from his or her card.

10. If the student has the correct answers on the card AND has shown that they were marked in response to the *correct questions,* then that student is the winner and the game is over. If the student does not have the correct answers on the card OR he or she marked the answers in response to *the wrong questions,* then the game continues until there is a proper winner.

11. If you want to play again, reshuffle the cards and begin again.

Have fun!

TERMS

aquifer	marble
atmosphere	metamorphic
avalanche	mineral(s)
basalt	Mohs scale
core	moon
crater	mountain
crust	ocean
earthquake(s)	Pangaea
equator	plain
erosion	plateau
eruption	pole
fault /Fault	Richter scale
fossil(s)	Ring of Fire
geology	river(s)
geyser	rocks
glacier	sedimentary
iceberg	seismometer
igneous	soil
latitude	tectonic plates
lava	tides
limestone	tremor
lithosphere	tsunami
longitude	volcano
magma	water cycle
mantle	weathering

Additional Terms

Choose as many terms as you would like and write them in the squares.
Repeat each as desired. Cut out the squares and randomly
distribute them to the class.
Instruct the students to place the square on the center space of their card.

Our Planet Earth Bingo

© Barbara M. Peller

Clues for Additional Terms

Write two or three clues for each new term.

_____	_____
1.	1.
2.	2.
3.	3.
_____	_____
1.	1.
2.	2.
3.	3.
_____	_____
1.	1.
2.	2.
3.	3.

aquifer	atmosphere
1. An ___ is an underground layer of porous rock, soil, sand, or gravel that yields groundwater. 2. An ___ stores and transmits useful quantities of water.	1. It is the mass of air that surrounds Earth. 2. The gaseous mass that surrounds a celestial body is its ___.
avalanche	**basalt**
1. The slide of a mass of snow down the side of a mountain is an ___. 2. A piece of falling ice or rock can cause a mass of loose snow to slide down the mountain. This is called an ___.	1. It is the most common type of solidified lava. 2. ___ is a dark gray to black igneous rock.
core	**crater**
1. The center of the Earth is called the ___. 2. Earth has a solid inner ___, probably made of an iron-nickel alloy. Its outer ___ is liquid.	1. The bowl-shaped depression at the top of a volcano is a volcanic ___. 2. An impact ___ is the result of an impact of a meteoroid or other projectile on a planet's surface.
crust	**earthquake(s)**
1. The outer layer Earth is called the ___. 2. Earth's ___ comprises the continents and ocean basins.	1. An ___ is the shaking and vibration at the surface of the earth. Most occur at fault zones. 2. The part of Earth's surface directly above the focus of an ___ is called its epicenter.
equator	**erosion**
1. The ___ divides the earth into the northern and southern hemispheres. 2. The ___ is a great circle. Every point on it is equally distant from the two poles.	1. The wearing away of the earth's surface by a natural process is called ___. 2. Most ___ is caused by running water. Other causes are glaciers, wind, and waves breaking against the coast.

Our Planet Earth Bingo

eruption 1. The sudden, violent discharge of steam and volcanic material is called an ___. 2. The bursting forth of lava from a volcano is called an ___.	**fault/Fault** 1. A ___ is a crack in the earth's crust resulting from the displacement of one side with respect to the other. 2. The San Andreas ___ is in California.
fossil(s) 1. A ___ is the plant or animal remains from a past geologic age. 2. Paleontology is the scientific study of ___.	**geology** 1. This science deals with the history of earth as recorded in its rocks. 2. Scientists who specialize in this field are called geologists.
geyser 1. A hot spring that ejects a column of water and steam at intervals is a ___. 2. Old Faithful in Yellowstone National Park is one.	**glacier** 1. A slow-moving mass of ice is a ___. 2. A ___ stays frozen from year to year.
iceberg 1. A floating mass of freshwater ice that has detached from a glacier is an ___. 2. Only about 1/8 of an ___ is usually seen above water.	**igneous** 1. When magma or lava solidifies, ___ rock is formed. 2. Granite and obsidian are ___ rocks.
latitude 1. ___ is measured from the equator. Positive values are north of the equator; negative values are south of the equator. 2. The equator is at 0° ___.	**lava** 1. When magma reaches the surface, it is called ___. 2. ___ is molten rock that comes out of a volcano or a fissure in the planet's surface.

limestone 1. ___ and shale are sedimentary rocks. 2. The calcite in ___ comes mostly from the remains of clams, corals and other organisms.	**lithosphere** 1. Earth's ___ includes the crust and the uppermost part of the mantle. 2. Earth's ___ is broken up into tectonic plates.
longitude 1. ___ is the angular distance east or west from the prime meridian. 2. Positive values of ___ are east of the prime meridian; negative values are west of the prime meridian.	**magma** 1. Molten rock beneath the earth's surface is called ___. 2. When ___ is emitted from a volcano, it is called lava.
mantle 1. The layer of earth that is beneath the crust is called the ___. 2. Earth's lithosphere is made up of the crust and the uppermost part of the ___.	**marble** 1. ___ is a metamorphic rock. 2. ___ is limestone that has undergone metamorphism.
metamorphic 1. Marble is a ___ rock. 2. ___ rock is rock that his been altered by exposure to heat, pressure and chemical actions.	**mineral(s)** 1. ___ that are cut and polished for use as ornaments are called gems. 2. Hardness, color and luster are three qualities of ___. The hardness of a ___ is classified according to the Mohs scale.
Mohs scale 1. The hardness of a mineral is classified according to the ___. 2. The ___ goes from 1 to 10. Talc measures 1 on the scale. Diamond is 10. Our Planet Earth Bingo	**moon** 1. Earth has one natural satellite, or ___. 2. Neil Armstrong was the first person to walk on Earth's ___.

mountain 1. A landmass that projects above its surroundings and is higher than a hill is a ___. 2. Mt. Everest is the highest ___ in the world. It is located in the Himalayas.	**ocean** 1. The body of water that covers more than 70% of the surface of the earth is called the ___. 2. The world ___ is divided into 5 separate bodies: the Pacific, the Atlantic, the Indian, the Arctic and the Southern.
Pangaea 1. We call the supercontinent that existed during the Mesozoic Era ___. 2. The single vast ocean that surrounded the supercontinent ___ is known as Panthalassa.	**plain** 1. A relatively level area with gentle slopes is called a ___. 2. A ___ is an expanse of level or rolling treeless country.
plateau 1. A ___ is a high, flat expanse of land. 2. A ___ is an extensive area with a flat surface that rises sharply above the adjacent land on one or more sides.	**pole** 1. Either end of Earth's axis is called a ___. 2. It refers to the region around either end of Earth's rotational axis.
Richter scale 1. It is used to describe the strength and duration of seismic waves. 2. An earthquake measuring 3 on the ___ is considered minor; one measuring 7 is strong; and one measuring 8 or more is great.	**Ring (Rim) of Fire** 1. Many earthquakes and volcanic eruptions occur in the Pacific ___. 2. ___ refers to the basin of the Pacific Ocean.
river(s) 1. A ___ is a natural stream of freshwater that flows toward an ocean, a lake, a sea or another river. 2. The Nile and the Amazon are the longest ___ in the world.	**rocks** 1. ___ are made up of minerals. 2. There are three main types: igneous, sedimentary and metamorphic.

sedimentary 1. Limestone, sandstone and shale are ___ rocks. 2. Some ___ rocks are formed by when loose particles which have been deposited on land or in water are pressed together.	**seismometer** 1. A ___ measures and detects seismic waves. 2. A seismograph that records actual earth movements is called a ___.
soil 1. ___ is made up of humus and disintegrated rock. 2. ___ refers to the upper layer of the earth that can be dug and plowed in order to grow plants.	**tectonic plates** 1. Earth's lithosphere (crust and upper mantle) are broken up into ___. 2. The location where two ___ meet is called a plate boundary.
tide(s) 1. The alternate rising and falling of the surface of the ocean are called ___. 2. There are two each day: high ___ and low ___.	**tremor** 1. The shaking or vibrating of the earth during an earthquake is called a ___. 2. A minor earthquake is sometimes called a ___.
tsunami 1. A ___ is a destructive sea wave caused by an underwater earthquake or volcanic eruption. 2. This great sea wave is sometimes wrongly called a tidal wave.	**volcano** 1. A ___ is a vent in Earth's surface through which molten rock and gases escape; the term also refers to the deposits that accumulate around it. 2. A ___ may be active, dormant, or extinct.
water cycle 1. The ___ is the movement of water between the atmosphere, ground and surface water bodies. 2. Some processes of the ___ include evaporation, condensation, and precipitation. Our Planet Earth Bingo	**weathering** 1. The decomposition of rocks, soil and their minerals through direct contact with Earth's atmosphere is called ___. 2. Mechanical ___ breaks rock into smaller pieces. Chemical ___ transforms rocks and minerals into different rocks and minerals. © Barbara M. Peller

Our Planet Earth Bingo

igneous	avalanche	lithosphere	volcano	seismometer
equator	latitude	tremor	minerals	marble
river(s)	Pangaea		limestone	moon
tectonic plates	atmosphere	iceberg	tide(s)	longitude
magma	tsunami	eruption	aquifer	lava

Our Planet Earth Bingo

volcano	sedimentary	mantle	ocean	magma
longitude	geyser	earthquake(s)	atmosphere	Ring of Fire
plateau	tsunami		fault/Fault	iceberg
minerals	pole	Pangaea	weathering	marble
lava	tremor	eruption	equator	aquifer

Our Planet Earth Bingo

volcano	iceberg	minerals	tide(s)	river(s)
tsunami	avalanche	basalt	latitude	Mohs scale
atmosphere	tremor		Richter scale	core
Pangaea	plateau	magma	geyser	mantle
aquifer	eruption	equator	weathering	lithosphere

Our Planet Earth Bingo

Pangaea	Richter scale	lithosphere	eruption	magma
metamorphic	geyser	latitude	ocean	river(s)
limestone	earthquake(s)		seismometer	tide(s)
iceberg	glacier	tremor	equator	basalt
aquifer	lava	mountain	geology	moon

© Barbara M. Peller

Our Planet Earth Bingo

lava	seismometer	atmosphere	earthquake(s)	eruption
metamorphic	iceberg	basalt	Pangaea	fossil(s)
sedimentary	moon		avalanche	lithosphere
marble	Richter scale	igneous	weathering	geology
minerals	equator	plain	fault/Fault	limestone

Our Planet Earth Bingo

core	Richter scale	mantle	sedimentary	moon
tide(s)	atmosphere	geology	latitude	river(s)
ocean	basalt		earthquake(s)	fault/Fault
equator	magma	weathering	mountain	limestone
longitude	iceberg	igneous	plain	lithosphere

Our Planet Earth Bingo

igneous	Richter scale	rocks	fossil(s)	minerals
longitude	lithosphere	tsunami	avalanche	river(s)
mantle	tide(s)		fault/Fault	crater
Pangaea	geyser	metamorphic	volcano	plateau
eruption	equator	weathering	mountain	basalt

Our Planet Earth Bingo: Card No. 7

Our Planet Earth Bingo

limestone	Richter scale	crust	tide(s)	crater
metamorphic	sedimentary	ocean	lithosphere	seismometer
river(s)	Ring of Fire		moon	earthquake(s)
aquifer	Pangaea	volcano	geology	geyser
tremor	equator	mountain	atmosphere	longitude

Our Planet Earth Bingo

fault/Fault	minerals	tsunami	river(s)	moon
geology	sedimentary	limestone	atmosphere	lithosphere
Mohs scale	igneous		avalanche	crust
crater	lava	magma	fossil(s)	rocks
geyser	weathering	core	volcano	seismometer

Our Planet Earth Bingo

tectonic plates	volcano	earthquake(s)	ocean	plain
moon	crater	latitude	avalanche	lithosphere
Richter scale	Ring of Fire		tide(s)	plateau
magma	marble	geology	weathering	Mohs scale
erosion	lava	mantle	longitude	limestone

Our Planet Earth Bingo

basalt	Ring of Fire	atmosphere	geology	longitude
crust	Mohs scale	fossil(s)	fault/Fault	latitude
metamorphic	sedimentary		mantle	tsunami
erosion	river(s)	weathering	equator	volcano
core	eruption	igneous	mountain	minerals

Our Planet Earth Bingo

minerals	geyser	Mohs scale	tide(s)	fault/Fault
tsunami	tremor	sedimentary	mountain	metamorphic
igneous	rocks		moon	ocean
eruption	seismometer	lithosphere	volcano	avalanche
Ring of Fire	crust	Richter scale	basalt	crater

Our Planet Earth Bingo

erosion	seismometer	core	Mohs scale	moon
sedimentary	crust	Richter scale	fault/Fault	plateau
tide(s)	earthquake(s)		tsunami	rocks
limestone	weathering	crater	Ring of Fire	volcano
equator	marble	mountain	igneous	fossil(s)

Our Planet Earth Bingo: Card No.13

Our Planet Earth Bingo

eruption	sedimentary	atmosphere	fault/Fault	erosion
crater	igneous	Mohs scale	avalanche	plateau
geology	tide(s)		mantle	earthquake(s)
marble	weathering	Richter scale	basalt	core
equator	ocean	Ring of Fire	longitude	limestone

Our Planet Earth Bingo

fossil(s)	fault/Fault	atmosphere	minerals	lithosphere
core	plain	latitude	sedimentary	geology
moon	igneous		river(s)	tide(s)
equator	Mohs scale	crust	weathering	erosion
longitude	geyser	mountain	mantle	tsunami

Our Planet Earth Bingo

earthquake(s)	water cycle	crust	plain	pole
ocean	Ring of Fire	rocks	metamorphic	tectonic plates
erosion	seismometer		moon	tsunami
Pangaea	geyser	equator	fossil(s)	volcano
geology	Mohs scale	mountain	crater	plateau

Our Planet Earth Bingo

erosion	soil	glacier	Mohs scale	equator
fossil(s)	geology	weathering	tide(s)	rocks
fault/Fault	tectonic plates		water cycle	crust
lava	longitude	limestone	atmosphere	plateau
magma	basalt	minerals	volcano	seismometer

© Barbara M. Peller

Our Planet Earth Bingo

lithosphere	Richter scale	crater	geology	ocean
lava	erosion	atmosphere	moon	basalt
fault/Fault	plateau		glacier	plain
Ring of Fire	latitude	weathering	tectonic plates	mantle
water cycle	Mohs scale	magma	soil	core

Our Planet Earth Bingo

moon	core	Mohs scale	crust	Ring of Fire
fossil(s)	eruption	plain	minerals	tectonic plates
soil	tide(s)		avalanche	atmosphere
mantle	water cycle	magma	geyser	glacier
river(s)	pole	longitude	limestone	mountain

Our Planet Earth Bingo

Ring of Fire	soil	tectonic plates	Mohs scale	avalanche
earthquake(s)	tsunami	metamorphic	magma	ocean
seismometer	rocks		Pangaea	glacier
lava	limestone	aquifer	geyser	water cycle
iceberg	tremor	pole	volcano	latitude

Our Planet Earth Bingo

core	lava	metamorphic	Mohs scale	marble
seismometer	glacier	crater	crust	igneous
plateau	longitude		soil	atmosphere
magma	minerals	water cycle	fossil(s)	limestone
Pangaea	pole	mountain	erosion	geyser

Our Planet Earth Bingo

river(s)	mantle	glacier	sedimentary	erosion
ocean	tectonic plates	lithosphere	crust	avalanche
crater	tide(s)		igneous	rocks
water cycle	lava	geyser	latitude	eruption
pole	basalt	soil	plateau	metamorphic

Our Planet Earth Bingo

earthquake(s)	soil	minerals	sedimentary	mountain
core	Ring of Fire	longitude	fossil(s)	latitude
mantle	erosion		aquifer	igneous
plateau	tremor	water cycle	basalt	geyser
marble	limestone	pole	magma	glacier

Our Planet Earth Bingo

earthquake(s)	Ring of Fire	eruption	soil	crust
moon	mountain	metamorphic	ocean	igneous
rocks	plain		erosion	plateau
marble	aquifer	water cycle	basalt	seismometer
iceberg	Pangaea	pole	tectonic plates	tremor

Our Planet Earth Bingo

Pangaea	metamorphic	soil	atmosphere	glacier
latitude	marble	fossil(s)	earthquake(s)	avalanche
seismometer	crust		aquifer	water cycle
plain	lava	tremor	pole	tectonic plates
mountain	eruption	crater	geology	iceberg

Our Planet Earth Bingo

glacier	soil	aquifer	ocean	plain
mantle	tide(s)	crust	Ring of Fire	earthquake(s)
marble	magma		tectonic plates	Pangaea
erosion	sedimentary	lava	pole	water cycle
rocks	geology	atmosphere	tremor	iceberg

Our Planet Earth Bingo

aquifer	crater	soil	Ring of Fire	tsunami
marble	mantle	fossil(s)	water cycle	avalanche
weathering	tremor		pole	Pangaea
plain	core	iceberg	metamorphic	latitude
erosion	tectonic plates	glacier	river(s)	rocks

Our Planet Earth Bingo

moon	Richter scale	volcano	soil	crater
tsunami	glacier	aquifer	magma	tectonic plates
tremor	plateau		plain	ocean
rocks	river(s)	longitude	pole	water cycle
sedimentary	fault/Fault	erosion	iceberg	marble

Our Planet Earth Bingo

glacier	Richter scale	plain	fossil(s)	fault/Fault
marble	magma	metamorphic	rocks	river(s)
seismometer	soil		avalanche	aquifer
tsunami	lava	lithosphere	pole	water cycle
earthquake(s)	crust	iceberg	core	tremor

Our Planet Earth Bingo

eruption	soil	ocean	fault/Fault	water cycle
latitude	plain	mantle	tectonic plates	avalanche
iceberg	tremor		rocks	metamorphic
marble	basalt	glacier	pole	aquifer
lava	minerals	core	Richter scale	lithosphere

www.ingramcontent.com/pod-product-compliance
Lightning Source LLC
Chambersburg PA
CBHW051419200326
41520CB00023B/7301